I0454583

THE EUCLID SPACE TELESCOPE

The Astonishing Discovery about EST & What You
Should Know; Astronomy in a New Light

By

Robin Flint

Copyright © Robin Flint, 2023

All rights reserved. No part of this publication may be reproduced, distributed, or transmitted in any form or by any means, including photocopying, recording, or other electronic or mechanical methods, without the prior written permission of the publisher, except in the case of brief quotations embodied in critical reviews and certain other noncommercial uses permitted by copyright law.

Table of Contents

Introduction

Dark energy and dark matter are mystic characters in the cosmos. These mysterious components of the Universe have eluded science since its inception and still leave us with several unanswered queries regarding our Universe.

However, in the pages of the cosmic chronicles there appeared a developing exploit that is the EST. The whole historic mission has changed the way in which we think about the universe and it reveals an entirely new era of exploration promising to clarify the enigmas of the universe.

ESA's 'Euclid' mission aims at unraveling the 'dark' Universe. The so-called 'unnoticeable' component of this mysterious universe accounts for over ninety five percent of all the energy out there as well as matter.

For a very long time, astronomers have sought to delve further into the luminous sources of the universe: for instance planets, stars, galaxies and gas. These elements though are just a tiny fraction of what the Universe constitutes. At first sight, it seems that 95 percent of the Universe consists of dark 'invisible' matter and energy.

According to scientists, this constitutes about one fourth of the universe and about seven-tenths dark energy. However, they can only be said to exist because they contribute towards influencing the development and propagation of the visible source without emitting, absorbing or reflecting any light; and for the moment scientists are yet to unmask them. The most fundamental question in the modern understanding of cosmology, therefore, is grasping their essence.

Today, ESA's Euclid space mission reveals its maximum specified full-range pictures of the universe. Never before has a telescope had the capability to make such surprisingly detailed cosmic pictures across this sort of extensive patch of the

sky, and searching up to now into the remote Universe. They show that the telescope is able to grow the widest three-D map of the Universe but, to find part of its mystery and uncover precise insights.

The discoveries by Euclid Space Telescope is a story of human ingenuity, dedication, and a relentless quest for understanding. From its inception as a visionary undertaking to its awareness as a state-of-the-art grandiose observatory, the EST has ended up a benchmark in technological development, supplying unprecedented insights into the fundamental building blocks of our universe.

This book is an odyssey through the awesome adventure of the Euclid Space Telescope, chronicling its inception, development, and the incredible discoveries which have reshaped the landscape of current cosmology. Through the lens of the EST, we withness the universe in a new light, analyzing cosmic anomalies with a degree of detail and clarity formerly too far.

Euclid Space Telescope (EST) and Its Mission

Euclid surfaced from two proposals that were submitted in response to the ESA Cosmic Vision 2015-2025 Call for Proposals, issued in March 2007: DUNE, the Dark Universe Explorer, and SPACE, the Spectroscopic All Sky Cosmic Discoverer. Both proposals suggested complementary ways to investigate dark energy, and after an evaluation study phase, they merged as a combined proposal: Euclid. In October 2011, Euclid was selected by ESA's Science Programme Committee for implementation, and it was officially adopted in June 2012.

The Euclid spacecraft is approximately 4.7 m high and 3.7 m in diameter. It consists of two main components: the service module and the payload module. The payload module includes a

1.2-m-diameter telescope and two scientific instruments: a visible-wavelength camera (the VISible instrument, VIS) and a near-infrared camera/spectrometer (the Near-Infrared Spectrometer and Photometer, NISP). Thermal control, propulsion, data processing electronics, telecommand and telemetry, electric power production and distribution, and attitude control are all included in the satellite systems found in the service module.

Launch and trajectory: Euclid is scheduled to launch from Cape Canaveral, Florida, USA, using a SpaceX Falcon 9 launch vehicle. Its operational trajectory will be a halo orbit around a point known as the Sun-Earth Lagrange point 2 (L2), at an average distance of 1.5 million km from Earth's orbit.

This unique site houses the Gaia and Webb space telescopes operated by ESA, and it tracks Earth's orbit around the Sun. Nominal mission duration is six years, with the possibility of extension (limited by the amount of cold gas used for propulsion).

Euclid is a cosmology survey mission, optimized to measure the proportions of dark energy and dark matter on cosmic scales. Euclid will take images in optical and near-infrared light; these images will cover more than one-third of the extragalactic sky outside the Milky Way, and depict billions of cosmic objects out to a distance where light has taken up to 10 billion years to reach us.

The image quality produced by Euclid will not be less than four times sharper compared to the results of sky surveys conducted on the ground. Euclid will also spectroscopically analyze thousands of billions of stars and galaxies in the near-infrared throughout the same sky.

This will allow scientists to investigate the chemical and kinematic properties of many objects in detail. Euclid will create a large archive of unique data, unprecedented in size for a space-based mission, enabling exploration across all disciplines in astronomy. Euclid's mass in orbit will be 2 tonnes (including 800 kg of payload module, an 850 kg service module, 40 kg of balancing mass and 210 kg of fuel).

Understand The Mysteries of Dark Energy and Dark Matter in the Universe

The Euclid Space Telescope's main goal is to better understand dark matter and dark energy, which together make up 95% of the cosmos. The reality of both remains entirely theoretical — although also necessary for scientists to construct a working understanding of the cosmos.

Dark energy is the name given to the unknown force causing the expansion of the cosmos to accelerate. Still, dark matter is invisible, its reality inferred from the motion of objects affected by its gravitational pull.

Dark Energy and Its Nature

One fact regarding the universe's expansion was rather certain in the early 1990s. The expansion would eventually slow down due to gravity, regardless of whether it had sufficient energy density to halt and collapse or an insufficient amount to cease forever.

Then came 1998 and the Hubble Space Telescope (HST) measurements of very distant supernovae showed that, a long time ago, the cosmos was actually expanding more slowly than it is now. So the expansion of the cosmos has not been decelerating due to gravity, as everyone assumed, it has been accelerating. No one anticipated this, no one knew how to explain it. But something was causing it.

Eventually scientists came up with three types of explanations. Perhaps it was a result of a long-discarded interpretation of Einstein's theory of gravity, one that contained what was called a "cosmological constant." Perhaps there was some strange kind of energy-fluid that filled space.

Perhaps there's something wrong with Einstein's theory of gravity and a new theory could include some kind of field that creates this cosmic acceleration. Scientists still do not know what the correct explanation is, but they've given the result a name. It's called "Dark Energy".

So, Dark Energy is the name given to the unknown force causing the expansion of the cosmos to accelerate. We know how important dark energy is because we know how it affects the cosmos's expansion. Other than that, it's a complete mystery. But it's an important mystery. It turns out that roughly 68% of the cosmos is dark energy.

Dark matter makes up about 27%. The rest—everything on Earth, everything ever observed with all of our instruments, all normal matter—adds up to less than 5% of the cosmos. Come to think of it, perhaps it should not be called "normal" matter at all, since it's such a small part of the cosmos.

One explanation for dark energy is that it's a property of space. Albert Einstein was the first person to realize that empty space isn't nothing. There are a lot of incredible properties in space, many of which we just begin to understand. The first property that Einstein discovered is that it's possible for more space to come into existence.

Also one interpretation of Einstein's gravity theory, the interpretation that contains a cosmological constant, makes a prediction: "empty space" can possess its own energy.

This energy would not be diminished as space expanded since it is a feature of space itself. More of this space energy would appear as more space is created. As a result, this form of energy would cause the cosmos to expand rapidly and rapidly.

Unfortunately, no one understands why the cosmological constant should indeed be there, much less why it would have exactly the right value to cause the observed acceleration of the cosmos.

Another explanation for dark energy is that it's a new kind of dynamical energy fluid or field, something that fills all of space but something whose effect on the expansion of the cosmos is the opposite of that of matter and normal energy.

Some scientists have named this "quintessence," after the fifth element of the Greek philosophers. But, if quintessence is the answer, we still do not know what it's like, what it interacts with, or why it exists. So the mystery continues.

A last possibility is that Einstein's proposition of gravity isn't correct. That would not only affect the expansion of the cosmos, but it would also affect the way that normal matter in galaxies and clusters of galaxies behaves.

This fact would give a way to decide if the solution to the dark energy problem is a new gravity proposition or not; we could observe how galaxies come together in clusters.

But if it does turn out that a new proposition of gravity is required, what kind of proposition would it be? How could it accurately describe the motion of the bodies in the Solar System, as Einstein's proposition is known to do, and still give us the different predictions for the cosmos that we need?

There are speculative propositions, but none are compelling. So the puzzle continues. The thing that's needed to decide between dark energy possibilities - a property of space, a new dynamic fluid, or a new proposition of gravity - is more data, better data.

Dark Matter and its Nature

No matter how conclusive different data appears to point in the direction of there being unseen matter, we still don't know what it is. Over the past half a century, astronomers have made numerous compliances that easily rule out a vast swathe of possible proffers of what dark matter is.

Dark matter can be created out of energy using huge flyspeck accelerators, in which subatomic particles are collided, generating energy that's used to produce brand-new particles.

The three possible forms proposed for dark matter are namely: Baryonic Dark Matter, Massive Compact Halo Objects (MACHOs), and Weakly Interacting Massive Particles (WIMPs). A MACHO is a proposed form of baryonic invisible matter.

The Baryonic Dark Matter

The most egregious possible form of dark matter isn't anything fantastic; rather, it's what's called baryonic dark matter, which just means that it's dark matter made of the familiar baryons, like protons and neutrons. There are a couple of clear options. The first is just that there exists a whole bunch of hydrogen gas out there, cold and unnoticeable, but tremendously massive.

And it was a possibility back when the new optical telescopes were being developed and dark matter was still new to us. Astronomers can now use giant radio telescopes to snoop on lots of effects that don't emit visible light. One of those effects is the radio emission of hydrogen gas.

And there's a lot of gas out there. Over the entire universe, there's a commodity like 10 times more mass in hydrogen gas than in all of the stars and planets. Hanging some figures on that, the glowing effects in planets make up about half a percent of the energy and matter of the universe. The bottom line is that we know about cool astral gas and it's not dark matter.

The Massive Compact Halo Objects (MACHOs)

The Massive Compact Halo Objects is the alternate egregious form of baryonic dark matter, or MACHOs for short. Any things that are dense, small, and invisible are MACHOs. These can be brown dwarfs or black holes. However, these objects could essentially go undetected across the universe.

Also there's the possibility of rogue planets, which are planets that have been thrown out of their solar system and now wander the universe, unattached to any star. You'd suppose that these objects would be hard to see, but astronomers are clever people. They realized that Einstein's proposition of general relativity would be helpful.

Einstein realized that this gesture had an intriguing corollary. Suppose you were looking at a distant star and another star passed directly in front of the distant star. The light from the far star would be bent as it traveled across the line of sight due to the near star's gravitational field.

The result would be that the distant star would appear as a ring or a halo around the near star. This is an illustration of gravitational lensing and the phenomenon is called an Einstein ring.

Still, the same phenomenon can be used to search for MACHOs. If a heavy, invisible large object passes in front of a distant star, it will appear that the star is getting brighter for a short while before

returning to its initial brightness as the intervening star moves on.

Regardless, in the 1990s, astronomers searched for instances of microlensing. And they did observe instances of microlensing with the appropriate

optical properties. They discovered that there are simply not enough MACHOs in the universe to explain dark matter after analyzing their data.

Weakly Interacting Massive Particles (WIMPs)

The third form of dark matter is Weakly Interacting Massive Particles, or WIMPs. WIMPs are particles that interact with each other and with normal matter only through gravity and the weak nuclear force. This means that they can pass through normal matter without being affected by it.

The idea of WIMPs was first proposed in the 1980s. Since then, astronomers have been searching for examples of WIMPs in the universe. They haven't found any yet, but they have found that WIMPs are the most likely form of dark matter.

Dark matter and its nature remain a mystery, but astronomers are getting closer to solving it. With the help of new technology and data analysis, they are able to rule out certain possibilities and narrow

down the search. Hopefully, one day soon, we will know what dark matter is and how it affects the universe

How to Identify Dark Matter

There are three experimental approaches that physicists have used to search for dark matter. They all assume that dark matter interacts with ordinary matter.

The first is that these particles might bump into ordinary matter particles, much like an electron might bump into another electron. That's one possible interaction. Since we would observe the result of a dark matter particle striking a matter particle directly, finding dark matter in this manner would be referred to as direct detection.

Alternatively, if dark matter is real and essentially a gas of particles that permeates the galaxy, then perhaps dark antimatter particles coexist with dark

matter particles. If they exist, then possibly a dark matter and dark antimatter particle might bump into one another and annihilate. Matter could potentially result from that interaction. This would be referred to as indirect detection if dark matter was found in this way.

And lastly, the third possible method for finding dark matter is by creating it out of energy. By using massive particle accelerators, we can make new particles out of the collisional energy of subatomic particles.

This specific approach of searching for dark matter is particularly appealing, as the detection and creation, and the entire process can be studied and designed meticulously.

EST's Development, Launch and Early Operations

Decades of exploration and technological development preceded the moment when the Euclid space telescope transmitted its first images back to Earth. Experimenters and masterminds from the Max Planck Institutes for Astronomy in Heidelberg and for Extraterrestrial Physics in Garching near Munich were involved.

They are part of the Euclid institute, which includes research institutes in 17 countries. Both the near-infrared camera (NISP, Near-Infrared Spectrometer and Photometer) and the optical camera (VIS, Visible Instrument) on the telescope were developed and constructed with their assistance.

Another team from the two Max Planck Institutes, together with collaborators from other institutions, now ensures the operation of the telescope and the logistics and quality of the data transmitted.

The Euclid spacecraft, built and operated by ESA, with contributions from NASA, lifted off from Cape Canaveral Space Force Station in Florida at 11:12 a.m. With the goal of determining the cause of the Universe's accelerating rate of expansion, EDT (17:12 CEST) began its purposeful mission on July 1, 2023.

Following launch and separation from the rocket, ESA's European Space Operations Centre received a signal from Euclid via the New Norcia ground station in Australia at 11:57 a.m. EDT (17:57 CEST).

An imaginary axis that passes through the Sun and Earth is where Euclid is located, behind the Earth. From that point on, a vast area of the sky is observed by the space telescope as it searches for other planets.

Then, the Earth's atmosphere cannot interfere, and Euclid has a clear view of the dark and mysterious universe. Euclid's Journey After its launch from Cape Canaveral, Florida, on July 1, the European Space Agency's new space mission. Euclid has transmitted its first test images to Earth, and they are spectacular and can shape the structure of the universe and learn more about the most distant objects in the universe.

Euclid traversed the vastness of space to reach its strategic vantage point nearly a million miles from Earth. The subsequent commissioning phase has been a testament to the telescope's cutting-edge capabilities, with all systems performing to expectation. NASA's Jet Propulsion Laboratory has been instrumental in contributing critical equipment for Euclid's instruments.

The establishment of a U.S.-based data center for Euclid will further integrate NASA-funded science teams into the international institute focused on unraveling the secrets of dark energy and dark matter. Euclid's images are astounding.

The VIS instrument comprises 36 CCD image detectors that capture 609 megapixels of image data, just under 17 megapixels per detector. Billions of visible light images will be seen in VIS. The NISP instrument has 16 detectors that capture 4.5 megapixels of image data, just under 300 kilo pixels per detector. NISP will be able to see millions of images in near-infrared.

The Groundbreaking Discovery made by EST

The macrocosm holds profound mysteries, chief among them the elusive dark matter and the confounding dark energy. In a landmark event, the Euclid mission, a testament to the joint efforts of the European Space Agency (ESA) and NASA, has unveiled its first batch of imagery on November 7.

They provide a fresh perspective on these cosmic mysteries. The mission's anticipated regular imaging operations are slated for early 2024, promising a revolutionary leap in our understanding of the macrocosm.

Euclid's first release of imagery presents a vibrant display of celestial wonders. The variety features a dense cluster of distant galaxies, intimate views of neighboring galaxies, the tightly-knit coterie of stars within a globular cluster, and the stellar cradle of a nebula.

The significant findings that the scientific community will eventually uncover are only hinted at by these images. The Horsehead Nebula was captured in the first batch of prints from the Euclid mission.

Euclid's mission, with a six-year timeline, aims to construct an unparalleled 3D map of the macrocosm. A third of the sky will be covered by this map, which will show billions of galaxies as far away as 10 billion light-years.

Euclid has a wide field of view to accomplish this. It can capture vast swaths of the sky very quickly. This is a different and distinct approach from telescopes like the James Webb Space Telescope, which focus on high-resolution images of smaller sky sections.

The telescope's wide survey is essential for studying dark energy — the mysterious force driving the accelerating expansion of the macrocosm.

By mapping the distribution of dark matter, which can only be detected through its gravitational influences, Euclid will help scientists discern the large-scale structure of the macrocosm and the role of dark energy over time.

Mike Seiffert, Euclid's project scientist at JPL, heralds the mission's first images as the dawn of a dedicated study of the dark macrocosm. The volume of record data that Euclid is expected to produce could help solve the puzzle of dark matter and dark energy.

Although the Roman telescope will cover a smaller sky section, its high-resolution images will provide a complementary analysis. Combined, they will help us peer ever deeper into the macrocosm's history. The partnership between Euclid and the Roman telescope, which is scheduled for launch by May

2027, will be a lynchpin in the comprehensive study of the macrocosm.

With the release of Euclid's images, the data is now available to researchers worldwide. There is great anticipation for the scientific studies that will result from these data. Euclid's data repository will grow as its mission advances. Scientists plan periodic releases to the global scientific community, hosted at ESA's European Space Astronomy Centre in Spain. Through the sharing of this information, astronomical data will become more accessible to a wider audience.

As mentioned previously, Euclid is a space observatory mission, primarily led by the European Space Agency (ESA). With significant contributions from NASA, the Euclid mission embarked on its journey to the macrocosm on July 1, departing from Cape Canaveral, Florida. It has since traveled nearly one million miles to position itself at a strategic point for its mission.

The Future of Space Astronomy in Cosmic Exploration

The Nancy Grace Roman Space Telescope project led by NASA will benefit greatly from Euclid's discoveries. The Roman mission will expand on Euclid's discoveries about the dark universe, and cast its attention beyond the solar system. Working as a team, they will embark on a quest to discover new planets and explore galaxies.

Euclid is poised to usher in a new era in cosmology as it approaches the start of regular science operations in 2024. It will give vital information about the existence and function of dark energy in the expansion of the universe by illuminating the cosmic web of dark matter and measuring the patterns of cosmic structures.

More about the dark universe

Majorly, the vast, invisible elements of the universe—dark matter and dark energy—are referred to as the "dark universe." Together, they comprise 95% of the total universe.

This greatly overshadows the 5% that is made up of the ordinary matter we can see and interact with. The structure and expansion of the universe are shaped by the dark universe, which is difficult to observe directly.

Dark matter is found by astronomers through its gravitational pull on light and galaxies. It influences the movement of stars within galaxies and links galaxies together. This implies the existence of a sizable mass that is neither reflexive nor emitting light. Comprehending dark matter is essential for tracking the development of cosmic structures and the history of the cosmos.

THE EUCLID SPACE TELESCOPE

Dark energy presents an even more profound mystery. It is the actual force responsible for the universe's accelerating expansion. Although its nature is unknown, it determines the universe's destiny. By studying fark energy, scientists hope to understand why the universe is growing at an increasing rate, a discovery that has profound implications for the ultimate fate of all cosmic entities.

Studying the dark universe depends on sophisticated observatories and telescopes like the Euclid mission, the Hubble Space Telescope, and the Roman mission. They assess the rate of cosmic expansion and describe the geographical distribution of dark matter, which are vital pieces of information for verifying theories regarding dark universe .

Conclusion

Exploring dark matter and dark energy in the cosmic universe has been an educative journey revealing intriguing secrets that the universe contains. Our journey has represented the insatiable desire of humankind to explore and create.

The revelation brought forth by the Euclid Space Telescope, a groundbreaking moment in our quest, has opened new chapters of cosmic understanding. It has allowed us to peer into the intricate cosmic tapestry in ways unimaginable before.

In the closing pages of "The Euclid Space Telescope: The Astonishing Discovery about EST & What You Should Know; Astronomy in a New Light," we find ourselves at a unique juncture in the vast cosmic tapestry we've explored together.

The journey has been one of wonder, discovery, and bottomless curiosity, where we have ventured into the depths of the unknown; I hope it has brought you into the heart of the Euclid Space

Telescope's astonishing revelations and the transformative impact of its discoveries. You've delved into the depths of the dark universe, challenged existing theories, and embraced new frontiers of possibility. Oh yes, please note that our voyage doesn't end here!

As we finally draw to a close our investigation into the incredible discoveries revealed by the Euclid Space Telescope, I would also be very interested in hearing your thoughts about this book. I value your reviews and feedback, as it will help me to better my work and supply you with more contents that interests you. Thanks for taking the reading.